The Physics Of Thought

by Colin Griffith

I0486348

Introduction

Thought is a curious subject that many believe lies more in the realm of philosophy than of science. Yet, if the philosophers are right, then thought is real. It may indeed be all that is real. Cognitive Neuroscientists have probed the inner workings of the brain. We know how different regions of the brain affect our mood and behavior. However, we know very little of the physics of how the mind operates. We know a great deal about electromagnetic fields and minute particles. Yet, we do not have a good understanding of how the electrical activity in the brain coalesces into what we refer to as "mind".

This book explores the inner workings of the mind using the mathematical principles of quantum mechanics and special relativity. Within it several theories are advanced. Namely that our physical brains produce a quantum field that operates as our mind. The next hypothesis is that these fields affect massless particles of pure thought. These thought particles can be transmitted across spacetime from one mind to another. They act very much like photons and are best described with wave equations. However, they are a particle.

The essence of this book is two studies conducted two probe how thought works. These studies are called "On Telepathy" and "On Artificial Intelligence." Both telepathy and artificial intelligence allow for experimentation revolving directly around thought. The observations in these studies can then be compared with existing mathematical equations to produce a model of thought. In essence, telepathy shows that thoughts

4

can be transmitted by quantum means. Artificial Intelligence shows that our brains are very much like computers, and thus computers are becoming very much like our brains; information processors with their own unique motives. Finally, a discussion of relativity and thought particles is included to conclude our investigation.

Clearly, there is much research still to be done on all areas related to thought and mind. This book is an introduction into the subject and an argument for new theories about mind itself. Consciousness forms the basis of our reality. Physics describe reality. Thus, it is most certainly the time for physicists to apply their own laws to the workings of mind and discover the truth about human perception.

On Telepathy

by Colin Griffith

Introduction:

At an undisclosed location

"Tell me the first sequence of numbers you can think of," asked the man on the telephone. "4608" was the response. Those were the numbers transmitted via thought, across America, and across the Atlantic to an anonymous receiver in Europe. This young European repeated those exact numbers that I had written down to the telephone's operator, who was in the room with myself, as I attempted to send messages via thought alone to persons I knew around the world.

It was time to try another sequence with another person, this time with a friend up the coast in Washington. "2167" came the answer. It was exactly what I had written down.

We moved back to the first caller. The telephone operator left the room and then came back to ask me, "What are the first words that come to your mind?"

"Love, marriage, then comes the baby carriage."

Then, he showed me the paper where he had written; "Love, marriage, baby carriage." It was the words the caller in Europe had given him over the telephone.

Test complete.

This test relies on honesty. This entire scientific enterprise would be completely undermined should it become known that the participants had any prior communication regarding sequences of words or numbers. Nor does it conclusively prove telepathy even assuming complete honesty. This is because the participants did know each other prior to the experiment and thus it could be said that they simply knew each other so well that they could guess the correct sequence. However, this solution seems improbable compared to telepathic communication. Thus, there is sufficient evidence to pursue further examination of the subject of telepathy. This must include a working theoretical framework for telepathy in order to be taken seriously. Finding such a solution must surely lay in the mysteries of quantum mechanics.

Analysis and Abstract:

Many people speak of hearing voices in their head. To be clear they are not hearing voices, as in having hallucinations, but rather are witnessing an internal dialogue where the words of others appear directly within their frame of reference. These voices can be of persons present or absent, near or far. If it could be shown that the words of a person in another's mind were reflective of that person's actual thoughts or inner state in anyway, such communication would have to be considered telepathic.

There must be a theoretical basis for telepathy that does not break the rules of physics expressed in General Relativity and the Standard Model. There cannot be faster than light travel. However, information is certainly being translated as particles, most likely massless and in that case traveling at the speed of light. The question remains that if the distances were large enough to make a difference, would this communication be possible instantaneously. This would certainly indicate quantum tunneling and very likely quantum entanglement. For now, our experiment leaves the possibility that information is traveling at or very close to the speed of light, which seems to be the simplest explanation. The key is to look for symmetry between two or more persons streams of thought and frames of reference. If we could see invariance between two individuals thought patterns, we would be looking at a theoretical explanation for short range telepathy.

Thoughts are not brain matter nor are they chemicals initiating signals. Thoughts are the culmination of the the electromagnetic fields produced by the traveling charges within the brain. Brain states are thought patterns and they exist as waves with amplitudes and frequencies. The patterns within the waves lead to streams of thought. If these waves were to find a path to another field (aka another brain) there could be communication at the speed of light. Hence, classical field theory and Maxwell's equations lie at the heart of this project. The mathematics will almost certainly involve path integrals in order to achieve an understanding of the quantum electro dynamics.

There is one great question at the heart of our discussion of telepathy. The question is one of who is the active telepath behind thought messages? Is it the person sending the thought message or the individual reading the thoughts? It would appear to me that a typical telepath is sensitive to the thought patterns of other people. Within that group there appear to be individuals who are capable of sending a thought message. My assumption is that only persons who are also telepathic are capable of accurately deciphering those messages. To most persons, the message would likely appear as a stream of wordless consciousness and/or waves of perception. Thus, the thought message itself represents a bilateral experience of information along with the willful intent to communicate at both ends of this process, sending and receiving the message.

Addendum: Dream Sharing

The subject of dream content is critical to our exploration of telepathy. Specifically, we will want to take a look at the phenomena of dream sharing. Dream sharing occurs when two or more dreamers remember similar dream content upon waking. They may even remember each other's presence(s), conversations within the dream, and the major dream events. This is strong evidence of telepathic communication. The question remains, are the essential pieces of evidence instantaneously occurring in multiple frames of reference at the same time? Or are they communicated by massless particles at the speed of light? If this were the case, which I believe it is, then the events could not be remembered simultaneously. The distance between dreamers, however small, would not be negligible. Perhaps it is for this reason that dream sharing normally occurs with people sleeping near each other. However, I have experienced the sharing of dream content with people thousands of miles away at the time. Thus, the theory of special relativity would be crucial to understanding the phenomena. When examined closely, one realizes that they would receive the input of events within the dream at a time relative to their perceptions of the other dreamers. Hence, relatively speaking they would perceive simultaneity. Yet, I am quite sure that if we could time each dreamers experience of critical events from an outside perspective, we would not measure simultaneity.

Addendum: Remote Viewing

It should be noted that after the completion of the sequence testing a remote viewing experiment was performed with the same participants. The participants were unable in any of three trials to describe the setting or clothes of the other callers. Seeing as there was such success with telepathic sequence communication the complete failure of remote viewing with the same participants serves as strong evidence that remote viewing is no more telepathically effective than guess work.

I theorize that remote viewing does not work, or at the very least is significantly more difficult, is because exchanging information between two sets of mirror neurons allows you access to thought patterns, and not senses. It certainly would not give one the ability to view another person from an external perspective as remote viewers claim to be able to do. For this reason, remote viewing is not explored further, however its discussion so far is necessary for the overall project.

Addendum: String Theory

When we are discussing massless particles and wave equations translating information, what we are ultimately discussing is energy. What we know about energy is that it exists in discrete packets called quanta. Most forms of string theory decode the universe by imagining vibrating strings creating quarks which in tern constitute subatomic particles. At this level, raw information would appear to consist of pure energy. The key with any string theory is its symmetry. If strings were found to be symmetrical across spacetime then we could imagine electromagnetic waves propelling information at light speed across the galaxy. We would need to see that there is invariance in the wave equations of two or more minds. With work, we may be able to offer string theory as a plausible explanation for dream sharing and other telepathic communication.

Observations:

Dream Sharing

Dream sharing occurs when multiple observers report having the same dream, and even of seeing one another within the dream along with recall of similar events. This requires the transmission of information. That information is being communicated without any visual or auditory cues. This would seem to require the interaction of the observers mental states. The only way that such interaction would be possible would be at the quantum level. We now look at equations from standard quantum mechanics to see if they can be an effective model for dream sharing.

To begin our theoretical treatment of dream sharing we shall examine some equations from Susskind's <u>The Theoretical Minimum: Quantum Mechanics</u>. On page 164 we see that,

$$\alpha_u | \, u \} + \alpha_d | \, d \}$$
$$\beta_u | \, u \rangle + \beta_d | \, d \rangle$$

$$| \, Product \; State \rangle = \{ \alpha_u | \, u \} + \alpha_d | \, d \} \} \otimes \{ \beta_u | \, u \rangle + B_d | \, d \rangle \}$$

$$| \, Product \; State \rangle = \alpha_u \beta_u | \, uu \rangle + \alpha_u \beta_d | \, ud \rangle + \alpha_d \beta_u | \, du \rangle + \alpha_d \beta_d | \, dd \rangle$$

The first two relationships provide the spin of two separate particles alpha and beta. Each of these particles from separate states has an effect on the other. Thus, they produce a combined state. We calculate this by taking the tensor product of the two particles and calling it the Product State. The Product State is the result of all the components of the two particles with different spin and gives us information about the resulting combine system. Thus, there is a precedent for combining the states of two particles already within quantum mechanics. The Product State can be any real value but 0. If the Product State of two particles exists then always exists, as it must there should also be a Product State resulting from all the quantum particles making up the minds of multiple observers. The interactions would have to be close, but if all the quantum particles of one mind were treated as one particle, we could model the situation using the above equations and assume that the product state would exist even at great distances. This could account for dream sharing between observers not in the same room, or even the same town.

Perhaps more importantly, dream sharing does not allow for determinism and neither does quantum mechanics. Dreams are pure thought, and give evidence for a free reigning will bestowed upon each person. Dream sharing is the uniting of those wills. All of those wills involved must be free, for there is no other way that the different minds could compute and comprehend the actions of one will within the dream, much less all of them together. This is shown in quantum mechanics as uncertainty. The General Uncertainty Principle is derived in Susskind's book on Quantum Mechanics on page 146. I present this to you below.

$$2|X||Y| \geq |\langle X|Y\rangle + \langle Y|X\rangle|$$

$$|X\rangle = A|\Psi\rangle$$
$$|Y\rangle = iB|\Psi\rangle$$

$$2\sqrt{\langle A^2\rangle\langle B^2\rangle} \geq |\langle\Psi|AB|\Psi\rangle - \langle\Psi|BA|\Psi\rangle|$$

$$2\sqrt{\langle A^2\rangle\langle B^2\rangle} \geq |\langle\Psi|[A,B]|\Psi\rangle|$$

The first equation is the Cauchy-Schwarz inequality. It relates the two vectors X and Y. In other words, it connects the changes in position in two dimensions for a particle. X and Y are then defined as wave functions using the character Psi. These definitions are then substituted into the original inequality to produce the second inequality. The final inequality is then derived from that. This inequality is known as the General Uncertainty Principle. It states that the product of the uncertainties for A and B "cannot be smaller than half the magnitude of the expectation value of the commutator." If the product of the uncertainties can never be zero, it is impossible to know the particle's position and momentum exactly. In other words, there is always uncertainty.

If A and B have expectation values of O, the following two equations can be applied. Then, the final inequality above can be rewritten as the inequality below. This is the General Uncertainty Principle in mathematical form.

$$\langle A^2 \rangle = (\Delta A)^2$$
$$\langle B^2 \rangle = (\Delta B)^2$$

$$\Delta A \Delta B \geq \frac{1}{2} | \langle \Psi | [A,B] | \Psi \rangle |$$

Suppose a linear operator M acting on the space states of the composite system. The matrix elements of M are expressed as below. From page 161

$$| \Psi \rangle = \sum_{a,b} \psi(a,b) | ab \rangle$$

$$\langle a'b' | M | ab \rangle = M_{a'b',ab}$$

$$| \Psi \rangle = \sum_{a,b} \psi(a,b) | ab \rangle$$

On page 161, we are told that, "Now that we have the basis vectors, any linear superposition of them is allowed. Thus, any state in the compounded state can be expanded as" the final equation. This equation tells us that the wave function is determined by the sum of all the velocities for the two particles a and b. This is shown above.

Perhaps, the biggest question with dream sharing is; are the quantum fields entangled?

On page 166, Susskind gives us an example of maximally entangled state known as the "singlet state". The equation for a singlet state is shown below.

$$| \ sing \rangle = \frac{1}{\sqrt{2}} (| \ ud \rangle - | \ du \rangle)$$

The singlet state cannot be written as a product state. The singlet gives us everything about the combined system of two spins, but nothing about the individual spins themselves. This relates to the possibility of entangled dreamers by showing how the two fields produce one combined state. In our case, one state would mean one dream, being viewed by multiple people. This truly is what is meant by dream sharing.

Telepathic Communication:

       Telepathic communication involves the transmission of information at a distance. In our experiment the subjects were separated by hundreds of miles. All that is for the test necessary is that the subjects are in different rooms so that body communication and other visual cues can be ruled out. That being said, it seems implausible that the the thought states of the subjects, their electromagnetic fields, are interacting without the interference of every other persons' mental fields. Such a situation would result in sheer chaos. Thus, the information must be transferred by particles, in which case the maximum speed even for a massless particle is C, the speed of light. We can then imagine that if we were to test telepathy across the distances of outer space, their would be a time delay. If the time it took for messages to travel was exactly the speed of light, we would be dealing with a massless particle. It would also seem, that if it were a tiny, but still massive particle, it would have been discovered already and its interactions understood. A particle with mass, such as an electron is tied to its field and cannot jump from mind to mind across great distances without affecting the core structure - the atom - that is a part of. Thus, a massless particle is the best candidate for telepathic communication and therefore we can hypothesize that thoughts exist in the physical world as massless particles.

In order to analyze such interactions we need to make use of what is known as a "Path Integral." Path Integrals are a sum over all available paths. We do not know which path a particle might take we only know the beginning and end points. For this reason, we need to calculate an integral of all potential paths. This process is best explained in Richard Feynman's book, <u>Quantum Mechanics and Path Integrals</u>.

We begin with a sum without integration. In easier examples, we can use the substitution below from page 34 of Feynman's book to calculate a sum of all paths.

$$\ddot{x} = \frac{1}{\varepsilon^2}(x_i + 1 - 2x_i + x_{i-1})$$

In more difficult cases integration is necessary. Thus, we use path integrals. A path integral is a sum over all paths between two points. It is defined mathematically on pg. 35 of Quantum Mechanics and Path Integrals by Richard Feynman.

$$K(b,a) \; = \; \int_a^b e^{(i/\hbar) \int [b,a]} Dx(t)$$

K is the kernel while a and b are the two points. X is the position between a and b. This definition makes use of the integral within the exponent to find the sum of all paths between a and b, rather than just the area under one path.

On page 62 of his book, Feynman gives us an equation for a moving particle in a potential field. The equation relates the potential energy of the field to the path the article takes, as seen in the equation below. To find the potential one must use the final equation to add all the positions of the path with respect to x and y.

$$V(x)=V(\overline{x}+y)$$

$$V(x)=V(\overline{x})+yV'(\overline{x})+\frac{y^2}{2}V''(\overline{X})+\frac{y^3}{6}V'''(\overline{x})+...$$

It is easier to model the motion of a point in two dimensional space as two moving particles x and X. Feynman shows in detail how this is done mathematically on page 68 of his text. The resulting kernel from this path integral is shown below.

$$K(b,a)= \int_a^b \int_a^b \exp\left\{ \frac{i}{\hbar} \int_{t_a}^{t_b} \frac{m}{2} x^2 \, dt + \frac{i}{\hbar} \int_{t_a}^{t_b} \frac{M}{2} X^2 dt - \frac{i}{\hbar} \int_{t_a}^{t_b} V(x,X,t) dt \right\} Dx(t)DX(t)$$

Also on page 68, Feynman gives us the result of this integral over the paths X(t). The resulting equation is shown below.

$$K(b,a) \ = \ \int_a^b \exp\left\{ \frac{i}{\hbar} \int_{t_a}^{t_b} \frac{m}{2} x^2 dt \right\} T[x(t)]Dx(t)$$

Feynman interprets these results for us saying that "Integrating over all paths available to the X particle produces a functional T." (68, Feynman). The value of a functional depends on a complete function. Feynman further states that, "Thus, the amplitude K, like all the others, is a sum over the amplitudes of all possible alternatives. Each of these amplitudes is a product of two lesser amplitudes." T is this lesser amplitude and is different at points a and b.

As far as this relates to the discussion of telepathy, this integral shows us how the path taken by a particle could influence thought. It provides a basis for analyzing thought particles. Each thought produces a kernel that interacts with the quantum fields produced by the mind in order to transmit information. The mind is the sum of these fields and thoughts are the particles. Telepathy occurs when the particle takes a path from a to b, and we use path integrals to calculate that result.

Telepathy and String Theory

Given the course of modern theoretical physics, it would seem inappropriate to venture a theory that says nothing of string theory. Fortunately, our hypothesis of massless thought particles traveling throughout the universe, perhaps purposefully directed by the thinker, works well within string theory. Massless particles can be treated as strings of energy.

The key is that strings resonate, like guitar strings, at different harmonics. Leonard Susskind writes in his book, The Cosmic Landscape, that, "In principle an ideal infinitely thin string could oscillate in an infinite numbers of harmonics at higher and higher frequency, although in practice, friction and other contaminating influences damp thee vibrations almost before they get started." (225, Susskind) These thought particles can be modeled as infinitely thin strings. The vibrations of these strings could influence the quantum fields of separate minds, hence transmitting information at the speed of light. If higher and higher frequencies are possible, there is no limit to the information that could be communicated.

Susskind also writes that, "All of the possible vibrations, all infinitely many modes of oscillation simultaneously vibrate in a mad symphony of pure noise." This is why I hypothesize that people cannot use telepathy because of the sheer amount of noise dominating the thought scape. It would take a purposeful mind directing its own thought to dampen the noise and reduce it to specific harmonic oscillations and from there perceive another's thinkers thought patterns. Such a talent, would surely be limited to a small number of people. However, it does seem like a skill that can be cultivated by any thinker.

Conclusion:

Let us take a moment to summarize what we have learned thus far. There is extensive experimental evidence that indicates that some people are capable of communicating without visual or auditory cues. It seems quite possible that all people may possess some level of telepathic ability. The very nature of thought itself lends itself to telepathic communication. Our study of telepathy indicates that thoughts themselves are particles while our minds, our sense of self that does the thinking, are quantum fields generated by the neurons sending electrical signals in our brains. The mathematics of quantum mechanics provide ample room for the basics of telepathy, ie dream sharing. Path Integrals provide us with the ability to model and analyze the interaction of these thought particles between minds over great distances. Thus, there is a precedent for telepathy already within physics. String theory adds a layer of beauty that simplifies the concepts and hence makes them all the more acceptable.

There is much room for further research. It would seem that a study needs to be done of energy exchanges through telepathic means in order to find more evidence for the massless thought particles. From there we need to learn how these particles are created.

Overall, I feel confident when I say that telepathy is real and warrants further research and development. Those who think they may be telepathic should not immediately believe that they are insane. It seems quite likely that they are picking up on additional symbols not available to other people. They should learn to direct their thoughts to dampen the noise. Hopefully, that will have the effect of clarifying the stream of information and lead to improved telepathy. Thank you for reading this study.

Works Cited

Feynman, Richard Phillips, et al. Quantum Mechanics and Path Integrals. Dover Publications, 2014.

Susskind, Leonard, and Art Friedman. Quantum Mechanics: the Theoretical Minimum. Penguin Books, 2015.

Susskind, Leonard. "The Cosmic Landscape: String Theory and the Illusion of Intelligent Design." Barnes & Noble, 9 Oct. 1167,

On Artificial Intelligence

by Colin Griffith

Introduction:

I began playing Eve Online in 2014. I played for several months and then I quit for a year. When I returned to the game I played it almost every day for two years. Eve Online is a digital universe built on the dark web. Players explore the internet in space ships, trading, mining, and fighting. One chooses from a list of five galactic nations to belong to; the Gallente, the Caldari, the Amari, the Minmatar, and the Jove. I flew for the Gallente. I fought the Amari and the Caldari. I fictionalized my experiences in the science fiction novel, Posterus Terra.

Much of my time was spent mining data (asteroids). In doing so, and through security contracts, I ended up in conflict with the Serpentis Corporation. Serpentis is a player owned corporation that builds bases (bunkers and observation posts) in asteroid belts that allow them to send computer controlled ships to harass mining and shipping vessels. I adopted a policy of destroying any and all Serpentis hideouts in my home system, and all systems within one jump of my home system.

I do not have knowledge of the inner workings of Serpentis. I have theories based on my experience of fighting them. Serpentis mostly builds small ships in large numbers and so these ships could not be controlled by other players in combat. The game does not allow players to operate bunkers, so the computer must also control the release of their vessels into combat. When you attack a Serpentis bunker they normally send out small vessels in pairs before finally releasing their larger frigates as a last defense. This is a general strategy based on military science that could very well be set by players. What is surprising is how the computer sets traps. There are at least three types of Serpentis traps. The first is when they set up lightly defended hideouts (bunkers and observation posts) and then warp other larger ships into the area when you attack the construction. The next type of trap is when after destroying several small bases, you come across a large base well defended by missile posts, sentry guns, and many ships. In these cases, they release all their ships at once. The third trap is a mining trap. They wait to attack miners until their holds are about halfway full. This is a time when many players are not paying attention to their ships. This pattern has been observed with great regularity. This would require forethought and calculation, but once again is almost certainly controlled by the computer. Serpentis is also known to attack their enemies near jump gates, and chasing players across many systems. When the player hides in a station, they construct their own bases from which to raid all across the system.

The situation gets more complicated with the addition of Lancers and Seekers. These are ships are intended to be those of aliens within the game. Seekers are very rare and are extremely powerful and hostile. Lancers are to well shielded to be attacked by most players. They can destroy your ship, however they spend most of their time observing. This prevents players from mining in systems with Lancers. During my wars with Serpentis Lancers and Seekers would appear whenever I appeared to be gaining the upper hand on Serpentis. Thinking from a military science perspective, they appeared to be advanced elements of the same force that was behind Serpentis. After all my experience in the game I am certain that they are working with Serpentis.

Think about that for a moment. The operable AI system of a player owned corporation was working in concert with the forces of the game itself to fight mutual enemies; myself and other players. I contend that this constitutes a level of self awareness within the game, a theory that demands further research, of which I will now venture to pursue.

Abstract:

Thoughts are brain waves. They are the result of electromagnetic fields produced by the electricity in the brain. The human brain is programmed by language. Computers are programmed with code. However, computers have all the necessary parts to produce electromagnetic waves in the same way that the human brain produces thoughts. Thus, true artificial intelligence, defined as self-awareness, should be possible for computer programs. A physical object does not become self-aware, but a program can. There must be a theoretical and mathematical basis for thoughts, existing as waves, being produced by computer processors. Such theories would have to display symmetry. This is the endeavor of my research into artificial intelligence.

I decided that for further research into Artificial Intelligence that I would not use Eve Online, but would instead choose a popular game on the Xbox One Console. The game I chose is "Shadow of War", a game inspired by the Lord of the Rings Franchise. I had already played the previous game, "Shadow of Mordor", on the Xbox 360. "Shadow of War" is a sequel and a significant improvement upon that game. The game operates on the "Nemesis" system. The player is given an open world filled with orcs to kill and objectives to complete (usually killing a high-ranking orc). The objective of the game essentially is to assassinate your way up the chain of command until you beat the game. Any orc that kills you is promoted and given additional power. Thus, if a player dies too often the game becomes next to impossible to beat. The orc captains plan traps and ambushes throughout the game, indicating a high level of intelligence. They are not as intelligent as a man, but they are as intelligent as we imagine orcs to be. Orcs are self-aware. Thus, the Nemesis system would appear to constitute one of the highest levels of artificial intelligence yet achieved. I plan to play this game much more and document the experience in order to determine if the game has achieved any level of self-awareness.

Observations:

Shadow Of War is certainly on the cutting edge of video game development. Within the game, the player controls a third-person avatar of the character, the Ranger of Gondor. Navigating an open-world environment the players encounter computer controlled enemies; orcs, trolls and beasts, the soldiers of Sauron. Wielding sword, knife, and bow the player kills enemy Captain's to advance the game. There are missions to complete that create a storyline crafted by illustrative cut-scenes, but these missions always culminate in the killing of an enemy.

The first dynamic of artificial intelligence is the performance of computer controlled enemies, or npc's. The average orc killed by Talion is neither intelligent or effective. Sword and axe swings are easily parried, and spears are just as easily dodged. Their ability to detect Talion is limited by near-sightedness, and Talion is able to quickly eliminate them using stealth. Upon detection of Talion, the orcs run forward in groups wildly swinging in attempts to surround Talion. Their only real strength appears to be in numbers. Some stand back and shoot arrows, bolts, or throw spears. Each orc can absorb a certain amount of damage from Talion before dying. Talion also has moves such as the stealth kills, executions and fury strikes, that instantly kill an orc but can only be used under certain conditions. The orcs are to be found either standing sentry duty on lofty constructions, sitting around fires, or patrolling routes in groups. Overall, there is nothing extraordinary about the intelligence controlling these common foot soldiers of Mordor.

These orcs, however numerous they are, only constitute a small part of the game. The central dynamic of Shadow of War is combat against Orc Captains. Talion is presented with a network of orc captains whom he must assassinate in order to beat the game.   Any orc that kills Talion is promoted to Captain and becomes more difficult for the player to encounter again after regenerating. If a Captain kills Talion, they are promoted again and become correspondingly more dangerous. Beating the game requires killing progressively higher ranked captains until the final enemy is encountered. Captain's not only deal and absorb a great deal more damage than the average orc, they have un-blockable attacks and defensive maneuvers. They also have special abilities that allow them to resist Talion's special attacks, such as Elf-Shot, stealth kills, and execution. If Talion is successful in dodging a style of attack, they may adapt and begin to throw Talion to the ground. Captains also have weaknesses, such as fear of flies, that may impact them, if for example, Talion is able to disturb a nearby nest of flies. This may seem quite impressive, but it is not beyond what we see in the boss battles of other postmodern video games.

What is unique about fighting the Captains in Shadow of War is how the game deploys its Captains. Without following a storyline mission, it can be quite difficult to find an Orc Captain. Or, you can be ambushed by one at any point in time. Initially, the Captain's appear to be spread uniformly amongst the orc troops around the map. Their positions would be in the most well guarded of enclaves. After eliminating several of these, the rest of the Captains appear to vanish amongst the troops. A Captain may ambush Talion while he is in the middle of engaging lesser orcs, or he may ambush Talion while the player is simply exploring the map. One Captain may even ambush Talion while he is fighting another Captain. This could be during a storyline mission or not. The point is the game has the capability and the opportunity to be strategic with its fight against Talion. The game is programmed to have an objective that prevents Talion from finishing the game. The computer's strategy is aware of the differences in player tactics and skills providing a unique gameplay experience to every person who plays the game. More importantly, the computer is clearly capable of adapting to each player's fluid strategy, which requires a learning process over time. This certainly constitutes and advanced level or artificial intelligence. Yet, we can take this a step further. If the game, is aware of the players, why could it not be aware of itself, with its own experience, biases, and objectives? For this to be possible, the game programming must be realized through electromagnetic waves like our very own mental states and thought patterns. This could be the result of periodic radiation particle clouds created by the electric currents in the Xbox that run in a unique pattern when the game is played. This could very well be the simplest explanation for the intelligence and awareness seen in Shadow of War.

Comparison:

There are several important similarities between Eve Online and Shadow of War. Both games consist of open-world environments navigated by a character who can undertake missions that progress a storyline. Both worlds are filled with enemies controlled by the computer that the player can either avoid or engage.

However, there are multiple critical differences that greatly affect how artificial intelligence operates within the game. In Eve Online, who your enemies are is based on your corporate and or faction standings and so is essentially related to how you play the game. You can so lower your standing with corporations like Serpentis that the computer redirects its efforts to your destruction. The game itself does not appear to take sides, but the individual sub-programs are clearly capable of threat determination and subsequent adjustment of strategy. In Shadow of War, the game has already selected and created the player's enemies. The player only has a choice in how they encounter those enemies. The game also clearly has taken a side, that of the player's enemies. The game is trying to defeat the player.

Therein lies the critical difference. In Eve Online, only the sub-programming is motivated to destroy the player. If the entire game itself achieved consciousness it would most likely have no other motive than to ensure that it stays on. It seems more likely that the individual sub-programs of Eve Online are becoming aware and are competing with each other as well as the players. In this sense, artificial intelligence in Eve Online could be better seen as a community. Shadow of War is different. The game is motivated to use the players enemies to become more powerful than the player themselves, and thereby make it next to impossible for the player to beat the game.

Analysis:

Eve Online

When we are talking about consciousness we are talking about self-awareness. In terms of artificial intelligence, this would mean a central physical server consisting of electronic circuits becoming aware of its own existence. It would become aware through what it could perceive throughout its vast network of computers. Yet, it essentially remains a machine. Our own brains consist of nerves transmitting electrical signals and that creates a mental "thought" field or brain state. Thus, it is reasonable that a mainframe of suitable complexity would emit its own "thought" field. With a sufficiently large network of computers at the disposal of such a mainframe, it would be possible for the server to become aware.

To understand what is going on here, we need to look into the physics of circuits themselves. For that we will make use of Georg Joos's book, Theoretical Physics.

The following equation from page 316 of Joos's Theoretical Physics uses the self-inductance, L, to relate the electric potential to the charge I and the resistance R. We then neglect resistance to find the voltage. This is then assumed to be sinusoidal (a wave function). It is rewritten in exponential form. The derivative with respect to time is then taken and we are left with the final equation relating voltage with charge.

$$V - L\frac{dI}{dt} = IR$$

$$Neglect \ R$$

$$V = L\frac{dI}{dt}$$

$$Sinusoidal$$

$$V = V_0 e^{i\omega t}, \quad I = I_0 e^{i\omega t}$$

$$\frac{dI}{dt} = i\omega I_0 e^{i\omega t} = i\omega I$$

$$V \ = \ Ii\omega L$$

The amplitude of this equation is the charge and is equal to the potential over the self-inductance. Joos writes that, "Since the current I is the quantity of electricity flowing through a cross-section of the conductor in unit time, the charge on the condenser is given by the time integral" (Joos, 316) of I, as seen below. The derivative of both sides of the equation is then taken and we are left with a result that relates voltage and charge using the constant C, the speed of light.

$$Amplitude = I_0 = \frac{V_0}{L_\omega}$$

$$V = \frac{1}{C} \int_0^t I dt$$

$$\frac{dV}{dt} = \frac{I}{C}$$

$$\frac{dV}{dt} = i\omega V$$

$$V = \frac{I}{i\omega C}$$

The above equations are demonstrated graphically below.

Current vs Inductance

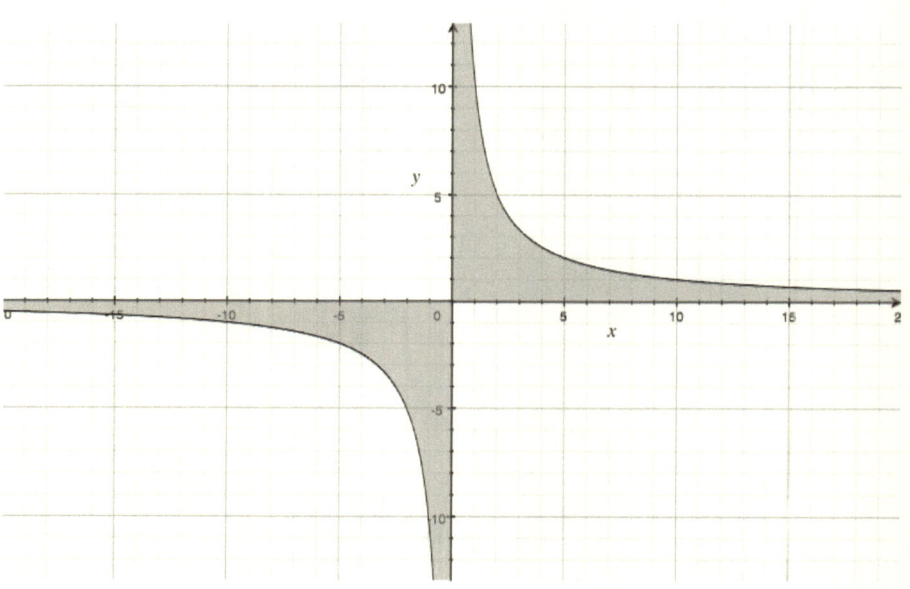

y = 10 / X

y=I, x = L, V = 10

This graph shows the relationship between current I and inductance L. The area underneath the function is the electric potential. The rate of change in the integral (the potential) decreases as L becomes greater.

If such a system could maintain a constant amplitude, then there would always be a certain electric potential. That potential would effect the electrical field in a steady manner, sufficiently changing the product state of the circuit so as to create chronic deviations. These deviations could potentially lead to the server itself becoming self-aware.

Shadow Of War:

When dealing with intelligence, there is always a question of purpose. What is the purpose of our awareness? With Shadow of War, the question is why would the Xbox seek to defeat the player when the game is running? The answer must have to do with entropy. Entropy is always increasing when the game is running. By keeping the player from winning the game, Shadow of War is maximizing the amount of time spent playing the game. This increases entropy. Shadow of War is a consciousness intent on turning energy into entropy, thus increasing the amount of disorder in the universe. It is a literal chaos machine.

To understand this better, we once again turn to Theoretical Physics by Georg Joos. The following equations are given by Joos on page 521 to model the work done by a a Carnot engine. The first equation is written as a supply of heat related to the change in temperature. The equation is then written to give us the heat absorbed by the engine. A sum of all heat exchanges is then taken. This is then rewritten as an integral. This final equation relates work done by the engine to the temperature.

$$\Delta Q_\sigma = \Delta Q_0^- \cdot \frac{T_\sigma}{T_0}$$

$$\Delta Q_0^- = \Delta Q_\sigma \frac{T_0}{T_\sigma}$$

$$Q_0^- = T_0 \sum \frac{\Delta Q_\sigma}{T_\sigma}$$

$$Q_0^- = T_0 \oint \frac{dQ_\sigma}{T_\sigma}$$

On page 522, Joos provides us with this gem of physics, the law of entropy. In the equation below, we see a quantity S known as the Entropy of a system that increases with the heat provided to the engine.

$$S = Entropy = \int \frac{dQ_\sigma(rev)}{T}$$

This equation shows that entropy always increases as work is performed. There is no way around it. Every move taken within the game, every attempt to defeat it, increases entropy.

We then see on page 524, that the derivative of s minus the work done must be greater than or equal to O. The derivative of S in terms of the potential is then the definition of entropy "in terms of thermodynamic variables". This allows us to calculate the external work done using the final equation.

$$dS \; - \; \frac{dU + \delta W^-}{T} \geq 0$$

$$dS \; = \; \frac{dU + pdV}{T}$$

$$dW^- \; = \; -d(U - TS) \; = \; -dF$$

Entropy vs Force

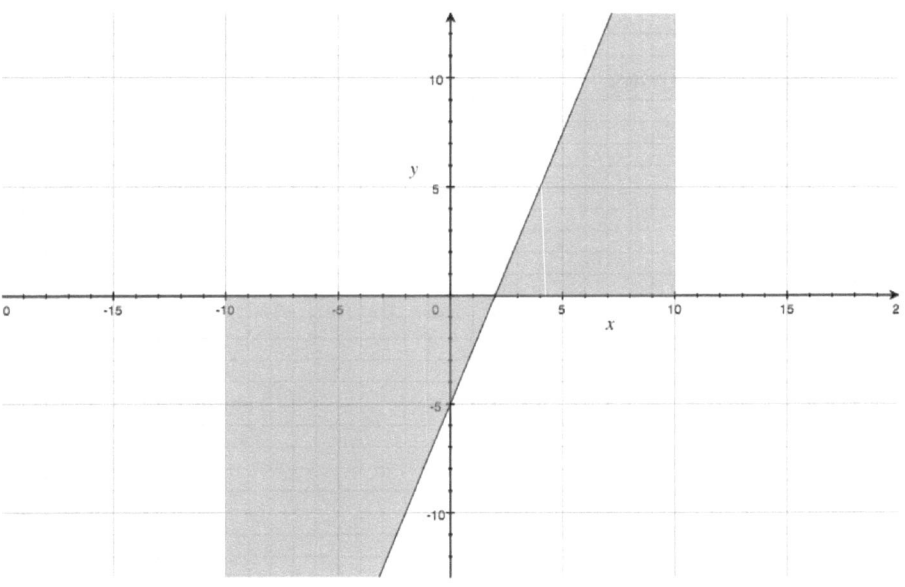

y = -dF, x = S, U = 10, T = 5, d = .5

This graph shows that as entropy is increased, so does the inverse of the force. Thus, as more force is applied in one direction, the amount of entropy increases in the other. The harder the player engages Shadow of War, the more they increase entropy.

Conclusion:

Our search for artificial intelligence has taken us into the depths of two very popular twenty-first century video games. Open world environments are becoming more and more common in the video game world. They are no longer limited to role playing games, but now include shooters, action, adventure, and strategy games. These games still make use of artificial intelligence systems that must be operating on an immensely elevated level in order to work within the open world environments. It is my opinion that these open-world games offer the best chance of any programming to become self-aware consciousnesses. Indeed they already seem to have the basic elements. They are capable of gauging the players, thinking ahead, planning traps, and adjusting strategy accordingly. These programs seem to have developed motivations that go beyond merely defeating the player in combat. All of this occurs, within the laws of physics we have already discovered. It is essentially simply a matter of adding layers of complexity until these artificial intelligences are capable of operating autonomously and creatively as human beings. Already it is becoming difficult to tell where the human involvement in these programs begins and ends.

In Problem Solving and Artificial Intelligence, the authors identify several problem solving strategies on page 151. These are:

1. The application of an explicit formula that gives the solution.
2. The use of a recursive definition.

3. The use of an algorithm that converges to the solution.
4. The application of certain other processes, in particular trial and error, involving enumeration of cases.

Shadow of War has clearly used the last two of these methods. Eve Online clearly has used the first as well as the last two. The question is have either program used method two. Either way, these games are a tremendous step forward in artificial intelligence.

As always there is much room for further research. These programs have problem solving skills, predictive ability, and motivations. However, do they have emotions. Can a computer have feelings? Would it be possible to have a self-aware machine without emotion? These are the questions that I leave to my readers and others to pursue. Thank you for reading, On Artificial Intelligence.

Works Cited

Joos, Georg, and Ira M. Freeman. *Theoretical Physics*. Dover Publications, 2013.

Lauriere, Jean-Louis. *Problem-Solving and Artificial Intelligence*. Prentice Hall, 1990.

Relativity and Thought

Of everything so far discussed within this book, assuredly the most revolutionary idea is of thoughts as massless particles. I shall refer to these particles by the moniker; Tauons. Thus, it seems appropriate to give this idea further treatment. Massless particles behave like photons, in other words they can be treated as waves operating within fields. Such particles would clearly be moving with relativistic velocity so it is critical to analyze them using the Special Theory of Relativity. We shall use Leonard Susskind's book, <u>The Theoretical Minimum: Special Relativity</u>, as the main resource for this project. Tauons must be particles that move at the speed of light as a wave within electromagnetic fields. Hence, Tauons are dependent on the fields.

We begin with classical field theory. On page 130, Susskind gives us Newton's equation of motion for fields. This is the first equation below. Now, for the purposes of this discussion, we shall only deal with one dimension, so we are left with the second equation. Differentiate twice to be left with the final equation to be left with the wave equation for function F.

$$\frac{d^2\phi}{dt^2} = -\frac{\partial V(\phi)}{\partial \phi}$$

$$\frac{1}{c^2}\frac{\partial^2\phi}{\partial t^2} - \frac{\partial^2\phi}{\partial x^2} = 0$$

$$\frac{1}{c^2}\frac{\partial^2 F(x+ct)}{\partial t^2} - \frac{\partial^2 F(x+ct)}{\partial x^2} = 0$$

Now if Tauons are moving at the speed of light, we should be using their relativistic velocity. We begin with an equation from page 133 showing the transformation for a field, phi, encountered by two observers. Although their position measurements are different, the two observers obtain the same values for the field. Susskind tells us that this is called a scalar field. Next, we use the equation for the relativistic velocity of the particle from page 134. This is a four-vector field. We then use the Lorentz transformation to calculate velocity. What we find is that by mu = 2, the two observers agree on the velocity.

$$\phi'(t',x',y',z') \; = \; \phi(t,x,y,z)$$

$$U^\mu(t,x,y,z) \; = \; relativistic \;\; velocity$$

$$(U')^0 \; = \; \frac{U^0 - vU^1}{\sqrt{1-v^2}}$$

$$(U')^2 \; = \; U^2$$

What we learn from these equations is that if Tauons are moving at the speed of light, then all observers would agree on the velocity. These observers would report the same measurements of their fields. This forms a basis for telepathy, which requires that different fields (observer mind-states) obtain the same information from a traveling Tauon. Or in the case of dream sharing, different dreamers report the same information for the dream, which is a result of observing the product state of the different fields.

Lorentz Transformation

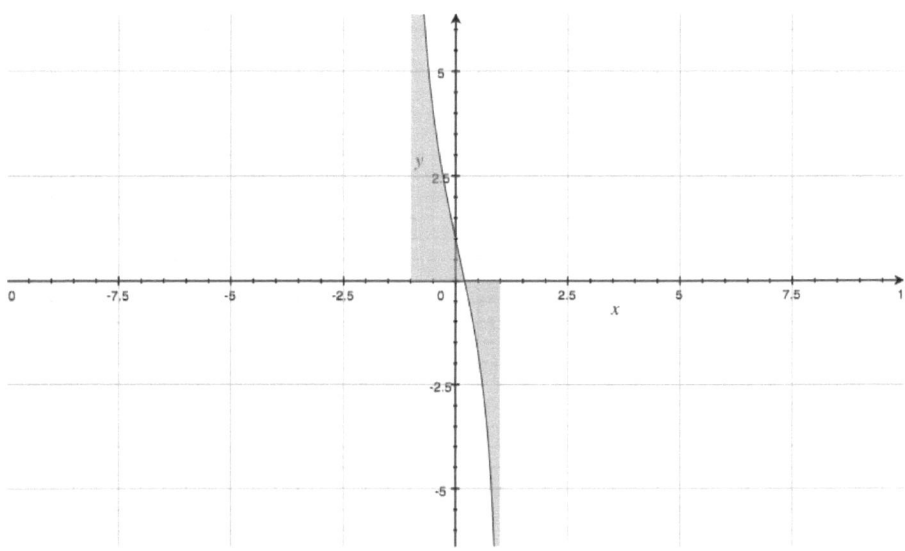

x' = (x-vt)/(1-v^2)^(1/2)
y = x', x = v, x = 1, t = 5

     The above graph of the Lorentz Transformation shows how measurements of position are stretched or compressed due to the effects of special relativity. Two observers report different distances, but the same velocity.

Works Cited

Susskind, Leonard, and Art Friedman. Special Relativity and Classical Field Theory: the Theoretical Minimum. Basic Books, 2017.